ALAN TURING AND THE POWER OF CURIOSITY

words by
Karla Valenti

pictures by
Annalisa Beghelli

with
Micaela Crespo Quesada, PhD

TO MARCO, MICAELA, AND ANNALISA—WITH PROFOUND GRATITUDE.

—KV

TO MY OWN TWO SUPER SCIENCE HEROES. MAY YOUR CURIOSITY TAKE YOU
WHEREVER YOU WANT TO GO!

—MCQ

TO MOMON, MY FAVORITE FIVE-HEADED MINION EVER! (EHI, DON'T TRANSLATE
IT... THESE ARE ONLY THE INITIALS OF CHILDREN IN MY LIFE!)

—AB

MARIE CURIE

Text © 2021 by Karla Valenti • Illustrations © 2021 by Annalisa Beghelli • Cover and internal design © 2021 by Sourcebooks • Internal images © Creative Commons 4.0 • Sourcebooks and the colophon are registered trademarks of Sourcebooks. • All rights reserved. • Marie Curie Alumni logo design is owned and licensed by the Marie Curie Alumni Association. • The characters and events portrayed in this book are fictitious or are used fictitiously. Apart from well-known historical figures, any similarity to real persons, living or dead, is purely coincidental and not intended by the author. • All brand names and product names used in this book are trademarks, registered trademarks, or trade names of their respective holders. Sourcebooks is not associated with any product or vendor in this book. • The full color art was created digitally. • Published by Sourcebooks eXplore, an imprint of Sourcebooks Kids • P.O. Box 4410, Naperville, Illinois 60567-4410 • (630) 961-3900 • sourcebookskids.com • Library of Congress Cataloging-in-Publication Data is on file with the publisher. • Source of Production: 1010 Printing Asia Limited, North Point, Hong Kong, China • Date of Production: November 2020 • Run Number: 5019874 • Printed and bound in China. • OGP 10 9 8 7 6 5 4 3 2 1

Deep beneath an icy mountain, in a dark and craggy cave...

Super Evil Nemesis was reading his new favorite book.

"It says here that if you want to take over the world, you must stop the spread of knowledge."

"Ooooh!" the minions said.

"That means no learning for anyone!" Nemesis went on.

"Ahhhhh!" The minions cheered and clapped.

"Let's get to work!" Nemesis slammed his book shut.

Meanwhile, in a city far away, one brave hero was about to be born, and he had a special power that would surely challenge Nemesis's evil plan.

And so begins our story:
an epic adventure to save the world!

One bright and sunny day in June, a very curious super science hero was born. He didn't know he was a super science hero, though everyone else figured it out pretty early on.

Including Super Evil Nemesis.

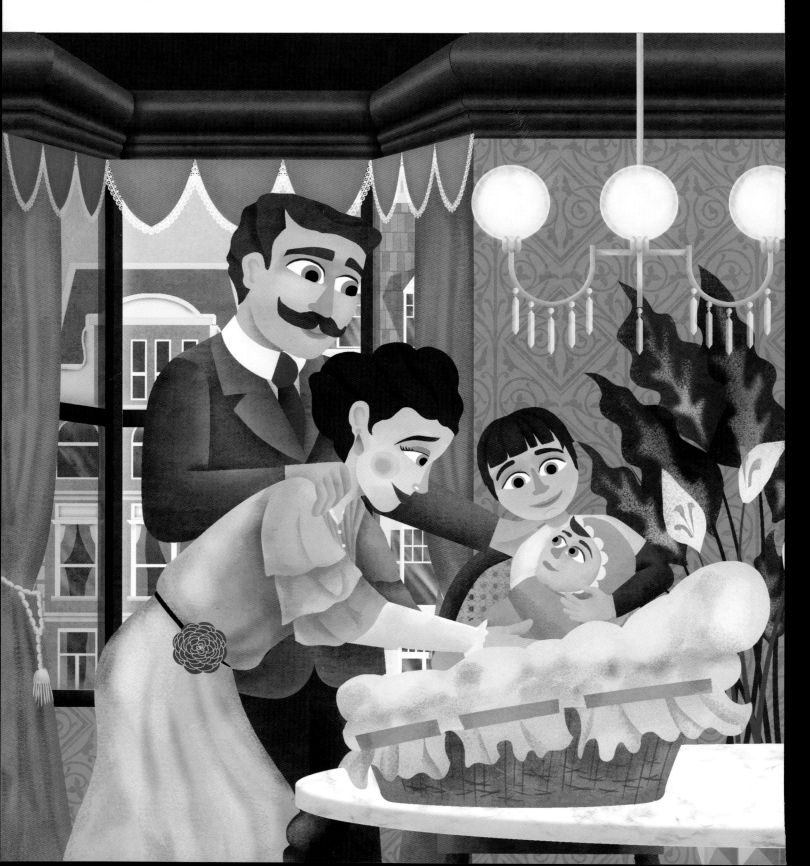

ALERT!

June 23, 1912—Alan Mathison Turing is born in Maida Vale, London, England. His father, Julius Mathison Turing is on leave from his position with the Indian Civil Service. His mother is Ethel Sara Turing. He has an older brother, John.

"We must confound and confuse him; make it impossible for him to understand anything. Eventually he'll be so frustrated that his curiosity will be extinguished!"

Nemesis called for one of his minions, Ms. Enigma.

"Without curiosity," he explained, "Alan will never want to learn a thing! His ignorance will make him easy to defeat. I'm counting on you to stop him!"

IGMX JZ TLCJYZJXF?

ALERT!

1936—Alan's power of curiosity is a remarkable force. He's published an important paper on the Entscheidungsproblem and has proposed new devices known as Turing Machines. They can solve any mathematical problem represented by an algorithm.

YOU MUST BEFUDDLE
AND CONFUSE HIM.
PREVENT ALAN
FROM SEEING
ANYTHING CLEARLY!

—NEMESIS

Super Evil Nemesis figured out that Entscheidungsproblem meant "decision problem" in German, but what was a Turing Machine, and whoever heard of an algorithm?

Things seemed to be getting out of hand.

FYL VLZX DQULNNKQ
MAN TYAULZQ GJV.
SCQWQAX MKMA
UCYV ZQQJAO
MAFXGJAO TKQMCKF!

—AQVQZJZ

ALERT!

September 3, 1939—Alan has been studying cryptology at Princeton University, and he has obtained a degree in mathematics as well.

PS—War has broken out across Europe! It's very scary.

The Daily Times
WAR
Germans invade!

BAD NEWS ABOUT PRINCETON, BUT GOOD NEWS ABOUT THE WAR! WAR CREATES A LOT OF CHAOS. ALAN WILL BE SO DISTRACTED THAT HIS CURIOSITY WILL BE QUASHED!

—NEMESIS

ENIGMA

ALERT!

September 4, 1939—Alan is not at all distracted. In fact, he's very focused on learning something new. Early this morning, he reported to Bletchley Park, the wartime station for the Government Code and Cypher School. He's going to help the British government decipher messages.

The German enemies had developed a very sophisticated system to communicate in code. They used devices called Enigma machines. Alan's task was to crack the code they were using. Fortunately, he had just the thing.

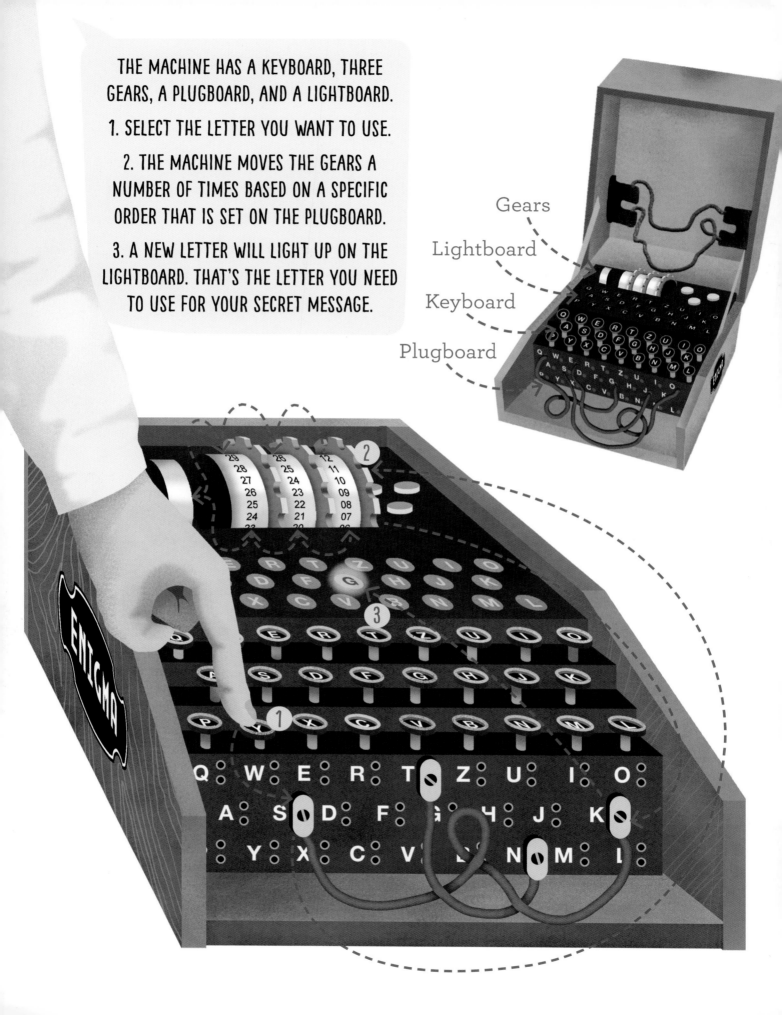

THE MACHINE HAS A KEYBOARD, THREE GEARS, A PLUGBOARD, AND A LIGHTBOARD.

1. SELECT THE LETTER YOU WANT TO USE.

2. THE MACHINE MOVES THE GEARS A NUMBER OF TIMES BASED ON A SPECIFIC ORDER THAT IS SET ON THE PLUGBOARD.

3. A NEW LETTER WILL LIGHT UP ON THE LIGHTBOARD. THAT'S THE LETTER YOU NEED TO USE FOR YOUR SECRET MESSAGE.

Gears

Lightboard

Keyboard

Plugboard

ENIGMA

"What now?" the scientists asked.

"There must be a word or a phrase used often, something that is said every time a message is sent," Alan continued. "If we can figure out what that word or phrase is, we can connect the coded letters to the uncoded letters. That will help us decrypt the rest of the messages."

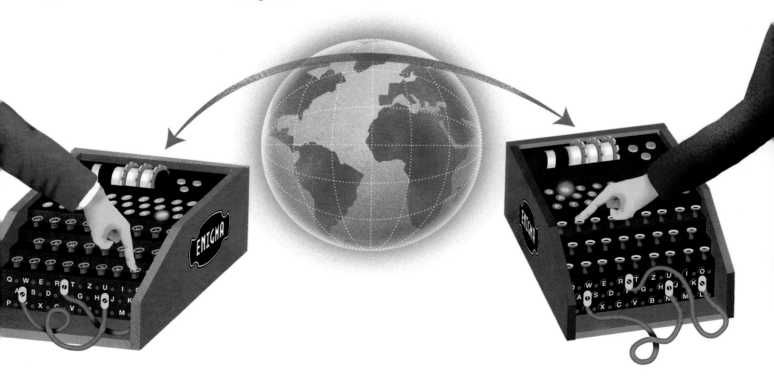

The team got to work straight away.

Alan had cracked the code!

"Now we just need to find a way to decipher the rest of the messages as quickly as possible," he told his team.

ALERT!

1947—Alan has written an important paper on "Computing Machinery and Intelligence." He thinks machines can learn and think for themselves! Isn't that amazing?

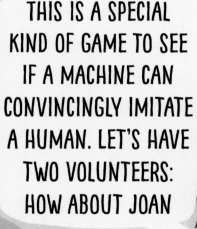

"Second question," Alan said. "Please add 12+15+23."

Alan reviewed the response. "'50' is correct!" Alan said. "Ms. Enigma likes math, and this problem was solved very quickly. I'm going to say it was Ms. E.

"Third question: What starts off with four legs, then has two legs, and ends with three legs?"

Alan looked at the final answer. "'A human,' that's right! And the answer probably came from Joan."

"Interesting!" Alan said. "I only guessed one out of three answers correctly. Ms. Enigma did a good job of imitating a human."

ALERT!

1951—Alan has cracked another code! He published a paper on "The Chemical Basis of Morphogenesis" where he says that stripes and spots are caused by two chemicals spreading through an organism at the same time. One chemical has instructions for making a color. The other chemical has instructions for turning off the color-maker. The instructions go on and off at different times, and that's what creates stripes or spots!

MORPHO-WHAT?!

Needless to say, Nemesis was not happy.

The same could not be said about the minions.

Alan's work would go on to be very influential to many Super Science Heroes, including you! He is considered the father of computer science and artificial intelligence. Without his curiosity, we wouldn't have any of these things:

As for Ms. Enigma, she would go on to wonder a great many things.

DELILAH, SHALL I TELL YOU A SECRET?
I LOVE SECRETS.
I AM A MACHINE.
YOU ARE?
BUT YOU PASSED THE IMITATION GAME! THAT MEANS YOU ARE AN INTELLIGENT MACHINE!
YES, I AM.

Nemesis, of course, was not so easily defeated.

"As long as there is ignorance in the world, I will prevail!"

MARIE CURIE AND THE POWER OF PERSISTENCE

GETTING YOUR WAY
AND OTHER SELFISH THINGS

BEING BAD: A MEMOIR

But as every member of the League of Super Science Heroes knows,

KNOWLEDGE IS POWER!

CODES AND CODE BREAKERS

COMPUTING MACHINERY AND INTELLIGENCE

PHILOSOPHIAE NATURALIS PRINCIPIA MATHEMATICA

CODES IN NATURE

COMPUTERS AND MACHINERY

MATHEMATICAL LOGIC

SCIENTIFIC THEORIES IN NATURE

RELATIVITY: THE SPECIAL AND THE GENERAL THEORY

Break Ms. Enigma's Code

Alan Turing cracked Ms. Enigma's code. If you would like to crack the code as well, use the following key.

This is an **Alberti disk**. The technique was first created in 1467 by architect Leon Battista Alberti. The device consists of two concentric circles—one has the standard alphabet, the other has the alphabet written out of order. To crack the code, you rotate the inner ring to match up with the letters on the outer disk.

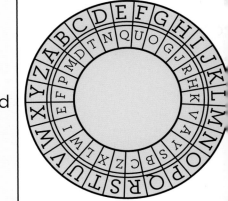

A	B	C	D	E	F	G	H	I	J	K	L	M	N	O	P	Q	R	S	T	U	V	W	X	Y	Z
M	D	T	N	Q	U	O	G	J	R	H	K	V	A	Y	S	B	C	Z	X	L	W	I	E	F	P

Now that you know how it works, try to decode this:

IGMX GMZ GMANZ DLX NYQZA'X TKMS? M TKYTH

There are many different kinds of codes. Test out your code-breaking skills with some of these other codes:

Simple Cypher

A simple cypher is based on substituting one character for another.

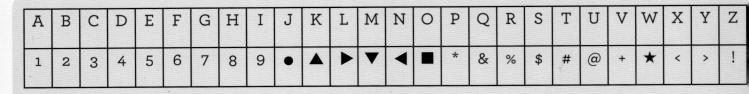

A	B	C	D	E	F	G	H	I	J	K	L	M	N	O	P	Q	R	S	T	U	V	W	X	Y	Z
1	2	3	4	5	6	7	8	9	●	▲	▶	▼	◀	■	*	&	%	$	#	@	+	★	<	>	!

Using this simple cypher, try to decode this message:

★81# 9$ 51$> #■ 75# 9◀#■ 2@# 496693@▶# #■ 75# ■@# ■6? #%■@2▶5

Caesar Shift

This code was named after Julius Caesar who used it to encode his military messages. The code is a simple cipher where you substitute each letter of the alphabet by shifting it right or left by a certain number of letters. It took codebreakers 800 years to learn how to crack it!

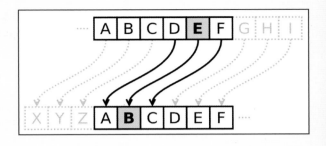

A	B	C	D	E	F	G	H	I	J	K	L	M	N	O	P	Q	R	S	T	U	V	W	X	Y	Z
									Rule for this example: shift the letters by 3															
D	E	F	G	H	I	J	K	L	M	N	O	P	Q	R	S	T	U	V	W	X	Y	Z	A	B	C

Using the Caesar Shift, try to decode this message:

ZKDW URRP GRHVQ'W KDYH GRRUV? D PXVKURRP

Morse Code

Named after Samuel F. B. Morse, an inventor of the telegraph, this code turns letters, numbers, and a few punctuation symbols into sequences of dots and dashes.

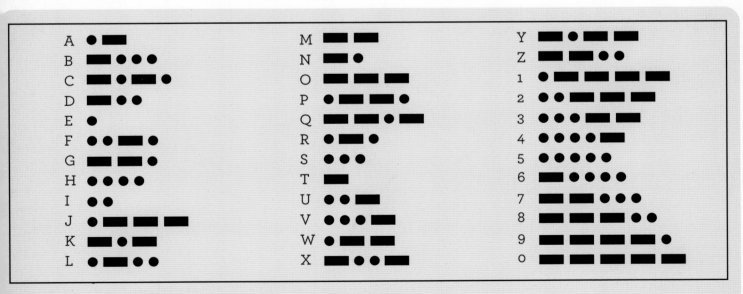

Using Morse Code, decode this message:

Other interesting codes include:

Binary Code

This code uses only two symbols (0 and 1) and is used by computers.

0	1	2	3	4	5	6	7	8	9	10
0	1	10	11	100	101	110	111	1000	1001	1010

Braille

This code was named after Louis Braille who lost his sight in a childhood accident. In 1824, he developed a code for the alphabet, where the characters have rectangular blocks with tiny bumps. The number and arrangement of these bumps distinguishes one character from another.

a/1	b/2	c/3	d/4	e/5	f/6	g/7	h/8	i/9	j/0
k	l	m	n	o	p	q	r	s	t
u	v	x	y	z					w

Pioneer Plaques

These are gold-aluminum plaques that were encoded with information about Earth and where our planet is in the universe. They were sent to outer space attached to the Pioneer 10 and 11 spacecraft as a message to other life forms.

GLOSSARY OF TERMS

ALGORITHMS

A process or set of rules to follow when making calculations or solving operations.

CRYPTOLOGY

The study of writing codes or solving them.

ENCRYPTION

The process of converting information into codes. Especially useful to prevent unauthorized access to information or to protect someone's privacy.

ENIGMA MACHINE

An encryption device designed to protect (encode) secret communication. It was extensively used by Nazi Germany during World War II.

IMITATION GAME

Developed as a test of a machine's ability to show intelligent behavior that is indistinguishable from human behavior. Also known as the "Turing Test."

MORPHOGENESIS

The origin and development of certain characteristics relating to the form or structure of things, like bodily tissues and organs.

ALAN TURING TIMELINE

Alan studies math and cryptology at Princeton University; obtains a PhD in math in 1938

Alan begins undergraduate studies at King's College, Cambridge

JUNE 23, 1912

Alan Turing is born

1926 | 1931–1934 | 1936 | 1936–1938 | 1939 | 1943

The first day of the term coincides with the 1926 General Strike in Britain. Alan is only thirteen, but he is so determined to get to school that he rides his bicycle 60 miles (97 km) to get there.

Alan begins research on "Delilah"

Alan publishes his paper "On Computable Numbers, with an Application to the Entscheidungsproblem"

He begins working with the Code and Cypher School, the British code-breaking organization

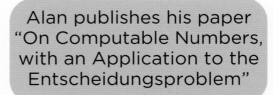

ENIGMA

Alan works on the design of the Automatic Computing Engine, presenting an important paper that is the first detailed design of a stored-program computer

He becomes Deputy Director of the Computing Machine Laboratory, working on software for one of the earliest computers

Alan begins studying mathematical biology and publishes "The Chemical Basis of Morphogenesis"

End of World War II

1945	1945–1947	1949	1950	1951–1952

JUNE 7, 1954:

Alan dies

He devises the "Turing Test" in a seminal paper on the topic of Artificial Intelligence, called "Computing Machinery and Intelligence"

MORE ABOUT ALAN TURING

Alan was a world-class long-distance runner, often running 40 miles (64 km) to London to attend meetings. He tried out for the 1948 British Olympic team but was unable to participate due to an injury (though his tryout time for the marathon was only 11 minutes slower than British silver medalist Thomas Richards's Olympic race time).

In 1946, he was appointed to the Office of the Order of the British Empire by King George VI for his wartime services, but his work remained a secret for many years.

Alan is attributed with creating the first computer chess program. In 2012, Russian chess grandmaster Garry Kasparov played against Turing's algorithm, beating it in 16 moves but recognizing it was an incredible achievement for the time.

GO BACK AND FIND

To put your curiosity to the test, Ms. E hid some items throughout this book. Can you find them all?

- A flower with petals that follow the Fibonacci Sequence. (It looks like this) ——

- Two spider webs reflecting the Golden Ratio. (They look like this) ——

- A sea serpent

- A piece of paper with incorrect coding

- Five ships on a map

- A joke written in code

- A clock missing its hands

- Mr. Opposition from *Marie Curie and the Power of Persistence*

- A creature with stripes that look like a tiger

- Books titled:

  - Marie Curie and the Power of Persistence

  - Relativity: The Special and the General Theory

  - Philosophiae Naturalis Principia Mathematica

  - Computing Machinery and Intelligence

  - Mechanical Intelligence

  - On Computable Numbers

  - Mathematical Logic

  - Morphogenesis

  - Intelligent Machinery: A Heretical Theory

HINTS

NFS

IGMX DQOJAZ IJXG

MECHANICAL INTELLIGENCE